Illustrations de: Sally Kindberg

Photographies: Peter Anderson, Paul Bricknell, Geoff Brightling, Jane Burton, Peter Chadwick, Andy Crawford, Geoff Dann, Mike Dunning, Neil Fletcher, Martin Foote, Steve Gorton, Frank Greenaway, Colin Keates, Dave King, Cyril Laubscher, Ray Moller, Tracy Morgan, Stephen Oliver, Susanna Price, Karl Shone, Steve Shott, Kim Taylor, Jerry Young

L'éditeur souhaite remercier:
Robert Harding Picture Library: Richard Pharaoh; Oxford Scientific Films: M.P. L. Fogden; Pictor International; The Stock Market:Photo Agency; Tony Stone Images: James Balog; Cosmo Condina; Zigy Kaluzny.

Doubles pages intérieures
Oxford Scientific Films: Rob Nunnington; Power Stock/Zefa: Tony Stone Images: Sylvestre Machado.

Jour
Ardea London Ltd: Ferrero-Labat; Colorofic: Michael Yamashita; Pictor International: The Stock Market: Photo Agency UK; Tony Stone Images; Gary Braash;
Mike McQueen; Laurence Monnerat.

Couverture
Robert Harding Picture Library: Richard Pharaoh; The Stock Market: Photo Agency UK house, Tony Stone

Sommaire

4   Le soir

6   La lune

8   Les animaux nocturnes

10 Le repos

Index Nuit

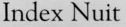

Animal 5, 8	Galago 9	Plume 4
Cactus 7	Gratte-ciel 5	Raton laveur 9
Chouette 4	Hôpital 11	Rêve 10
Chauve-souris 8	Infirmière 11	Soir 4, 5
Ciel 6, 7, 11	Insecte 8	Soleil 5, 6
Comète 7	Jumelles 6	Sommeil 5, 10
Corps 10	Lit 10	Ver luisant 9
Coucher du soleil 4	Lumière 5, 9	Ville 5, 11
Crépuscule 4	Lune 5, 6, 7	Yeux 8, 9
Escargot 8	Médecin 11	
Espace 7	Noir 5, 9, 11	
Étoile 6	Nourriture 4	
Feu d'artifice 11	Oreille 8	
Film 11	Partenaire 9	
Fleur 7	Planète 7	

NUIT

Découvre le monde à la lueur de la nuit

Claire Llewellyn

Les éditions Scholastic

Le soir est le début d'une longue nuit noire.

Chasseur de nuit
Au crépuscule, les chouettes quittent leur perchoir et partent à la chasse pour chercher leur nourriture.

Les plumes souples des ailes ne font pas de bruit.

Le crépuscule

Lorsque la nuit tombe, le ciel devient de plus en plus noir et la Lune apparaît. L'air se rafraîchit.

Anecdote nuit
Le soir, le Soleil se couche à l'ouest. C'est ce qu'on appelle le crépuscule.

Anecdote jour
Le matin, le Soleil se lève à l'est. C'est ce qu'on appelle l'aube.

Bonne nuit !

Le soir, la plupart des animaux sont prêts pour une bonne nuit de sommeil.

Les lumières de la ville

Les lumières s'allument lorsque vient la nuit. Les gratte-ciel s'illuminent dans le noir.

La Lune et les étoiles éclairent la nuit.

Demi-lune

Croissant de lune

La Lune, cette coquette
La Lune semble différente chaque nuit. C'est parce que le Soleil en éclaire différentes parties. Croissant de lune.

Les jumelles permettent de voir les étoiles lointaines.

Comètes

Parfois, des boules de glace brillantes appelées comètes traversent le ciel dans la nuit.

Anecdote nuit
La Lune est une grosse boule de roche qui flotte dans l'espace. Elle tourne autour de notre planète, la Terre.

Anecdote jour
Le Soleil est l'étoile la plus proche de la nôtre. C'est une immense boule de gaz en feu.

Une comète a une longue queue composée de gaz et de poussières.

On peut voir les pétales blancs la nuit.

Fleur de la nuit

Ce cactus ouvre ses fleurs blanches et odorantes la nuit.

Les animaux nocturnes ont des oreilles pointues et de grands yeux.

Elle entend tout !
La chauve-souris chasse en écoutant l'écho qui rebondit sur les insectes en vol.

Le paradis des escargots
Les escargots sortent la nuit pour se nourrir, lorsque l'air est plus frais et plus humide.

Une abeille peut-elle se nourrir la nuit?

Non. L'abeille se nourrit dans les fleurs colorées qu'elle ne peut voir que le jour.

Soulève les rabats pour comparer la vie le jour et la vie la nuit.

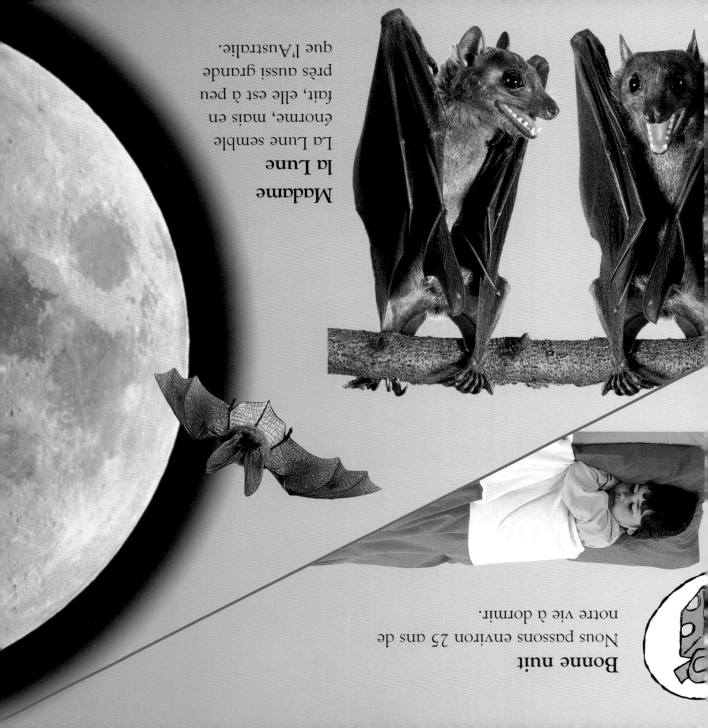

Madame la Lune
La Lune semble énorme, mais en fait, elle est à peu près aussi grande que l'Australie.

Bonne nuit
Nous passons environ 25 ans de notre vie à dormir.

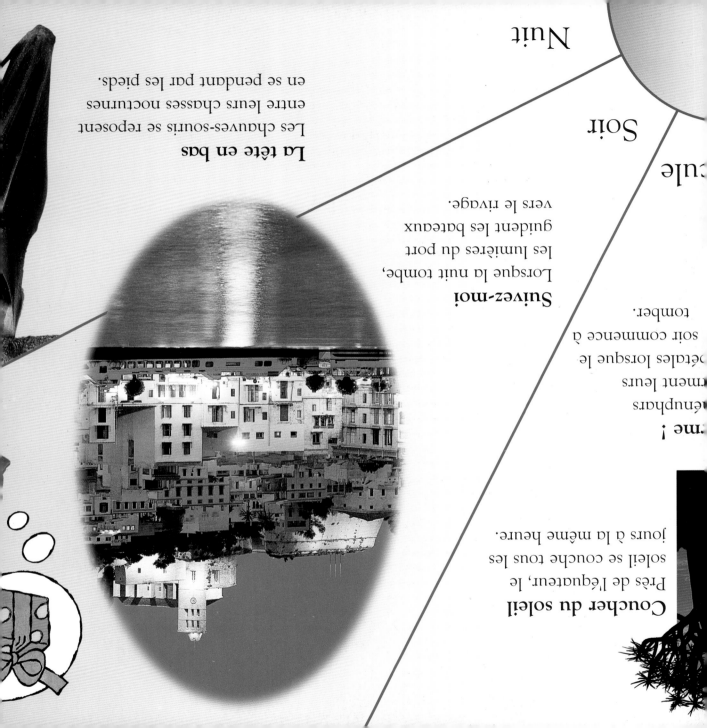

Nuit

La tête en bas
Les chauves-souris se reposent entre leurs chasses nocturnes en se pendant par les pieds.

Soir

Suivez-moi
Lorsque la nuit tombe, les lumières du port guident les bateaux vers le rivage.

...pétales lorsque le soir commence à tomber.

Coucher du soleil
Près de l'équateur, le soleil se couche tous les jours à la même heure.

Anecdote nuit
Les feux d'artifice illuminent le ciel étoilé à la nuit tombée. C'est un beau spectacle.

Les gens s'allongent pour dormir.

Ils travaillent la nuit
Les hôpitaux sont toujours occupés. Les médecins, les infirmiers et les infirmières travaillent toute la nuit.

Anecdote jour
Les parades de rue, avec leurs costumes multicolores, leurs drapeaux et leurs chars sont une vraie merveille.

Vie nocturne
La ville ne manque pas d'intérêt la nuit. On peut aller voir jouer une pièce de théâtre ou encore aller au cinéma.

The Little Golden Book of HOLIDAYS

By Jean Lewis
Illustrated by Kathy Wilburn

A GOLDEN BOOK • NEW YORK
Western Publishing Company, Inc., Racine, Wisconsin

Copyright © 1985 by Western Publishing Company, Inc. Illustrations copyright © 1985 by Kathy Wilburn. All rights reserved. Printed in the U.S.A. No part of this book may be reproduced or copied in any form without written permission from the publisher. GOLDEN®, GOLDEN & DESIGN®, A GOLDEN BOOK®, and A LITTLE GOLDEN BOOK® are trademarks of Western Publishing Company, Inc. Library of Congress Catalog Card Number: 83-82199
ISBN 0-307-03115-2/ISBN 0-307-60239-7 (lib. bdg.)
G H I J

June hung up her calendar for the new year. It was a present from Rob, her friend next door.

There was a page for each month. On every page, June found special days she could look forward to, holidays to celebrate. There was a whole year of fun!

"Happy New Year, Rosy!" June said, hugging her cat. Then June put on her coat and snow boots and ran next door to thank Rob for the gift.

Snow was still on the ground in February. But June's kitchen was warm when she and her mother baked special Valentine's Day cookies.

"I think I'll surprise Rob," June said when the cookies were all done. She wrote Rob's name in chocolate icing on a few cookies.

Then she left them in a box outside Rob's door, rang the bell, and ran away.

When his doorbell rang, Rob was cutting a valentine out of red paper.

"I'll bet June made these," he said, tasting one of the cookies.

When he finished cutting out the paper heart, Rob wrote June's name on it with glitter.

He slipped the valentine under June's door and ran home to finish the cookies.

There were special cookies at school a few days later. They looked like little cherry trees.

"Guess whose birthday is coming," said Mrs. Noonan, the teacher.

"George Washington's!" shouted the class.

"That's right," said Mrs. Noonan. "George Washington, our first president."

"We celebrate another president's birthday on Presidents Day," Mrs. Noonan said. "Who can help us guess his name?"

June went to the board and drew a log cabin. "Abraham Lincoln!" everyone cried.

"Yes," said Mrs. Noonan. "He was our sixteenth president." She handed out construction paper so the whole class could make special stovepipe hats to wear. They looked just like the one that Abraham Lincoln wore.

When June turned her calendar to April, she started to think about making another kind of hat. She and her mother went to the store and chose a plain white bonnet. Then they bought ribbons and yellow cloth.

When they were done cutting and sewing, June's Easter bonnet was covered with yellow flowers. "Just like the ones in our window box!" June said.

The day before Easter, June and Rob went to a big Easter egg hunt in the park. They found colored eggs in the tall grass.

Rob found the most eggs. His prize was a big chocolate bunny.

On the way home, Rob's mother told them why the egg is an important part of Easter. "It means new life," she said. "That's what Easter is all about."

In church on Easter Sunday, June and her parents heard all about the meaning of Easter. They heard about the new life that came to the world on that special Sunday long ago.

The church was filled with white and yellow flowers. June loved the organ music, and the singing, and the sunlight streaming in through the windows.

The sunlight grew stronger and warmer as time went by. Before long it was time for a July Fourth picnic.

June's family met Rob's family in the park. Together they spent all of Independence Day running races, climbing trees, and playing games. They ate a picnic supper.

The fireworks started just after sunset. The sky was lit up red, white, and blue. There was even a fiery sign that said, "Happy Birthday, America!"

The leaves turned from green to red to gold. June and her mother went shopping for pumpkins and found one that was just right.

June's mother scooped out the seeds. Then she helped June carve a mischievous grin in the shell. When they had cut out eyes and a nose and put a candle inside, June had her very own jack-o-lantern for Halloween.

June worked on her costume for a week. On Halloween night, she looked like a scary witch.

Rob was scary, too. He was dressed as a ghost.

Rob's father took June, Rob, and their friends around the neighborhood. "Trick or treat!" the children called when the neighbors opened their doors. They came home with sacks full of candy.

There was an announcement at school the next day. "Thanksgiving is coming!" said Mr. Green, the children's teacher. "Our class is going to put on a play about the first Thanksgiving."

Each child had a part to play. Rob played a Pilgrim. June played an Indian. In the play, Rob invited June and the other Indians to share the Pilgrims' harvest feast.

Early Thanksgiving morning, June's father took June and Rob to watch the big balloons being blown up for the parade.

Their favorite balloon was the giant turkey, but it needed more air in its tail feathers before it was ready to go.

When June and her father got home, they helped Mother prepare a big turkey of their own. Grandmother and Grandfather came for dinner, and Aunt Jane and Uncle Bill, and June's little cousins Michael and Elizabeth.

They all sat down together and said a prayer of thanks. Then they ate roast turkey, stuffing, cranberries, yams, and mince pie.

Rosy liked the turkey best.

A few days later, June and her mother started their Christmas shopping. They went to the city's biggest department store.

June chose a warm scarf for Dad, a mouse on wheels for Rosy, and a new calendar for Rob.

Then she went upstairs to see Santa. She told him what *she* wanted for Christmas.

On Christmas Eve, June and Rob and their parents and friends gathered around a big Christmas tree in the park. They sang "Silent Night" and "Jingle Bells" and "We Wish You a Merry Christmas." One of the neighbors brought candy canes for everyone.

On Christmas morning, there were presents under the tree at June's house. June found three storybooks, a doll, and a paint set.

"Santa brought everything I asked for," June said happily.

June's parents liked their presents, too. Rosy played with her new mouse all day long.

But they did not forget that Christmas is more than just presents. June's father read them the story of the special baby who was born in a manger such a long time ago.

A week later, June helped Rob hang up his new calendar. Together they looked at all the special days coming up, and they remembered all of last year's fun.

"Soon we can hunt for Easter eggs again," said June.

"And watch the fireworks on July Fourth," said Rob.

"Happy New Year!" they cried at once. And they ran outside to play in the snow.